Abel Hernández-Muñoz

RODENTS, CUTE PETS

Abel Hernández-Muñoz

RODENTS, CUTE PETS

Approach to the breeding and care of these friendly pet mammals.

ScienciaScripts

Imprint

Any brand names and product names mentioned in this book are subject to trademark, brand or patent protection and are trademarks or registered trademarks of their respective holders. The use of brand names, product names, common names, trade names, product descriptions etc. even without a particular marking in this work is in no way to be construed to mean that such names may be regarded as unrestricted in respect of trademark and brand protection legislation and could thus be used by anyone.

Cover image: www.ingimage.com

This book is a translation from the original published under ISBN 978-613-9-43744-3.

Publisher:
Sciencia Scripts
is a trademark of
Dodo Books Indian Ocean Ltd. and OmniScriptum S.R.L publishing group

120 High Road, East Finchley, London, N2 9ED, United Kingdom
Str. Armeneasca 28/1, office 1, Chisinau MD-2012, Republic of Moldova, Europe
Printed at: see last page
ISBN: 978-620-8-10409-2

RODENTS, CUTE PETS

APPROACH TO THE BREEDING AND CARE OF THESE FRIENDLY MAMMALS

MSC. ABEL HERNÁNDEZ MUÑOZ

INDEX

1. INTRODUCTION

Rodents, the generic name for certain mammals, whose main characteristic is their dentition: rodents have a single pair of incisors in each jaw; these are broad, curved or semicircular, have a sharp, chisel-like edge at the end and are used by the animal to gnaw. The front surface of each incisor is made up of hard enamel, while the back surface is made up of soft dentine, which is the area that is worn away when the animal gnaws, so that this wear keeps the edge chiselled and cutting. This is related to the presence of open cavities in the pulp of the tooth, which results in the continuous growth of the incisors and therefore the need for continuous wear of the tip of the incisors. Rodents do not have canines and there is a space (the diastema) between the incisors and molars. The jaw joint is arranged in such a way that the incisors can be positioned either forwards to gnaw or backwards to allow the molars to chew. Both the lips and the incisors form a mechanism of very diverse usefulness; they are not only used to collect food, but also to build nests or dig burrows. In addition, most rodents are also characterised by well-developed ears.Rodents are the order with the most species within the mammalian group; there are more than 400 genera and about 2 000 species. They are adapted to live in all types of terrestrial and freshwater habitats, as there are no marine rodents, and are distributed throughout the world, as humans also introduced them to places where they did not live naturally, such as New Zealand, some oceanic islands and other sub-Antarctic islands. Rodents are divided into three major groups: a first group contains seven families and includes among others squirrels, woodchucks and beavers, rats and mice, and kangaroo rats (suborder Scyuromorphs); a second group comprises about nine families, of which two contain most of the species: these are the New World rats and mice, and the Old World rats and mice (suborder Myomorphs); finally, the third group contains about fourteen families, almost all South American, and includes among others porcupines, capybaras, agoutis and chinchillas (suborder Histricomorphs). The largest rodent of all is the capybara, the rest of the species are generally small. They are animals very prolific and some species can have several litters in a single year. Some species are aquatic, others are terrestrial and live in burrows dug in the ground. There are also arboreal species and some 35 species, the so-called flying squirrels, have semi-arboreal habits. On the other hand, some rodents are serious pests for crops and grain stores. Other species, such as the grey rat and the black rat, are involved in disease transmission. Muskrats and beavers are valued for their fur and for building dams that help prevent soil erosion. Albino varieties of rats and mice have been of great importance for scientific research as laboratory animals, while gerbils and guinea pigs are well known pets throughout the world.

Scientific classification: Rodents constitute a single order: the order Rodentia or Rodents.

Pets play an important role in households around the world. Your personal pet ownership experience will be most enjoyable if you carefully consider which pet will best suit your family, your home and your lifestyle. Frustrated expectations are a major cause of people abandoning their pets, so make an informed decision. Take your time, include your family, and pay careful attention to the following questions if you are considering a pet rodent.

SELECTION OF A PET RODENT

What is so special about rodents?

Rodents such as hamsters, mice, gerbils and guinea pigs are examples of "pocket pets", so called because they are very small and require less care than other types of pets. For these reasons, they make excellent first pets for young children.

What options do you have?

Hamsters

The most common hamsters are Syrian hamsters or golden hamsters, but there are also albino hamsters (white with pink eyes). Hamsters kept in pairs or groups may fight, so they should generally be kept alone.

Gerbils

Similar in size to hamsters, gerbils are more active and social. Unlike hamsters, gerbils are happiest in pairs or small groups. Potential owners should be aware that in some countries it is illegal to buy and keep gerbils as pets.

Mice

Although mice can be docile and playful, they are a little more nervous than hamsters or gerbils. Females have no problems when They are found in pairs or small groups, but usually the males fight with each other. The mice most commonly found in pet shops are albinos, but there are also "fancy" mice that come in a variety of colours.

Rats

Rats are sociable and thrive in same-sex pairs. They are larger and easier to handle than some smaller rodents, rarely bite, and often form a very close bond with their owners. Rats come in a wide variety of colours and require a larger cage and more attention than smaller rodents.

Guinea pigs or guinea pigs

They are the largest rodents kept as pets, and their size and gentle temperament make guinea pigs very popular. They are sociable, unlikely to bite, and are comfortable being in same-sex pairs. They can also be noisier than other rodents.

What are some characteristics of rodents?

• Compared to dogs and cats, rodents have a shorter life expectancy. Young children should be aware of this so that the "sudden death" of their pet is not so frightening and upsetting for them. The average life expectancy is 2 to 3 years for hamsters and gerbils, 1 to 3 years for mice, 2 to 4 years for rats, and 5 to 7 years for guinea pigs.

• Housing is a critical component of keeping your rodent healthy and safe. Rats and guinea pigs require larger cages than those generally sold in pet shops. All rodents should have adequate space to move and exercise. Exercise wheels can be wonderful sources of activity and stimulation for small rodents. Cages should have secure latches, as rodents can be excellent escape artists. Secure housing is particularly important if your family has other pets. If you allow your pet to be out of its cage, keep an eye on it at all times.

• Rodents love to nibble! It is important to give your rodent chewing accessories to preserve its physical and mental well-being.

• Guinea pigs require more time and attention from their owners. They have more demanding dietary needs than other rodents. For example, they require fresh hay and vegetables. They also require vitamin C supplements, as they do not produce vitamin C on their own and must obtain it in their diet. Long-coated guinea pigs require frequent brushing on a regular basis to prevent tangling of their fur.

Who will look after your rodent?

As the owner, you will be responsible for your rodent's food, shelter, companionship, exercise, and physical and mental health throughout its life. Although children should be involved in pet care, it is illogical to expect them to be solely responsible. An adult must be willing, able and able to make the time available to supervise their care.

Does a rodent fit into your lifestyle?

Because they can live in cages, rodents can easily be kept in flats, condominiums and houses. Although rodents require les maintenance than other pets, that doesn't mean you shouldn't commit to giving them time and care. Rodents are famous for their ability to have numerous offspring. Buying and breeding a rodent for the sole purpose of having children witness the birth process does not demonstrate responsibility as a pet owner. If your female rodent becomes pregnant, it is your responsibility to find good homes for the offspring. In addition, some females may injure or eat their own offspring and this can be traumatic for children. To avoid the problem of accidental breeding, do not mix males and females in the same cage. Hamsters, mice and rats are nocturnal animals, which means they are most active at night. This can be frustrating for children, as they will want to play with their pet during their normal sleeping period. In addition, the noise that the rodent produces during the night as it moves around in its cage may disturb some people's sleep. Hamsters, mice and rats are often difficult to wake during the day and are likely to become cranky if disturbed. Since gerbils are most active during the day, they are the best choice for a child's schedule. Guinea pigs can be active during the day or at night. Keep in mind that if there are children in the family, they should be instructed in the proper handling of their pets; for example, rodents should NEVER be lifted by the tail!

Can you keep a rodent?

Although rodents can be adopted or purchased relatively cheaply, you should anticipate additional costs for housing, food, accessories and veterinary care during the lifetime of your pet.

Where can you get a rodent?

Most rodents are purchased from pet shops. You can also purchase rodents from reputable breeders, rescue groups and animal shelters. Always inquire about the return policy in case you discover that your pet is unhealthy.

What should I look for in a healthy rodent?

Avoid animals that appear to be sick. Trying to nurse a sick rodent back to health after you have purchased it rarely works. If your potential pet's living environment is dirty, smells bad or looks suspicious, do not buy or adopt any animal from that place. A healthy rodent should have no discharge from its eyes, nose or muzzle. The animal should look active and try to flee and resist handling to some extent, but should not panic when handled. There should be no evidence of coughing, sneezing or panting. Be sure to examine the tail area of the animal. It should be dry and free of diarrhoea or hardened faeces. This is an especially important thing to check when buying or adopting a small hamster; baby hamsters can have a disease called "wet tail", which can be fatal.

How should you prepare for a rodent?

Make sure your pet's cage has fresh litter, food and water and plenty of space for exercise (e.g. wheels for suitable species). All rodents need materials to build their nests. Guinea pigs need dark or enclosed areas for sleeping.

A veterinarian should examine all rodents within 48 hours of acquisition. This physical examination is critical to detect signs of disease and to help new owners learn about proper care. Since many problems are caused by misinformation and improper care, the first visit to the veterinarian will help prevent owners, however well-intentioned, from making mistakes that may contribute to the premature death of the animal. Not only is the veterinarian qualified to assess the health of your new companion, but he or she can also advise you on parasite control, nutrition, neutering, socialisation, training, grooming and other care that may be necessary to ensure your pet's well-being. Your veterinarian should continue to examine your rodent at least once a year to detect any health problems that may arise. By diagnosing and treating a disease early, there is a greater chance of a lower cost and a favourable outcome.

When you acquire a pet

You accept responsibility for the health and well-being of another living being. You are also responsible for the impact your pet may have on your family, friends and community. A pet will be a part of your life for many years to come. Invest the time and effort necessary to make the years you spend together a happy one. When you choose a pet, you make a commitment to care for it for a lifetime. Choose wisely, keep your promise, and enjoy one of life's most rewarding experiences!

• Litter is an important part of your pet's care. To avoid odours in the cage, use absorbent litter and change it regularly. Some types of litter should be avoided as they may be toxic to small animals; consult your veterinarian before choosing.

• Rodents should always be handled calmly and with slow movements, and a safe "refuge" area should be provided in their cage.

• To allow your pet to spend time out of its cage in a controlled manner, build a safe, easy-to-clean area for it to roam, such as a desk, large container or run. Leashes can be used with large rodents.

• Before purchasing your rodent, consult a veterinarian who is familiar with these species and join a club or group to learn more from experienced owners.

MOUSE BREEDING

Important:

To start with, you need at least 2 or 3 females, a male and a cage. 2.- Once the mice have reproduced, and it is time to separate the offspring, you will need 2 more cages to separate the youngest from the parents. 3.- A space is needed to place the cages and a certain dedication of time.

HOW TO BUILD THE CAGE

• The cage can be made of different materials, but the easiest is a plastic box of 50cm x 40 cm and 30 cm high (it can be the same or higher), and it will be enough to accommodate the 3 females and one male.

• We will have to cut the lid (as in the photo), and fix a piece of wire mesh with a maximum hole of 0.5 cm x 0.5 cm (it can be a smaller hole, but not bigger).

• On one side, and at a height not too high from the base, a hole should be made in the plastic to pass the drinker through, and two more to fix it on the outside with a wire.

MAINTENANCE

• To absorb urine etc., a layer of about 3 cm of wood shavings and/or straw (which can be purchased from agricultural cooperatives) is needed, as well as small pieces of paper or cardboard for nest construction and shelter. **This litter should be renewed every week, with the exception of the breeding season**, when it is better not to disturb with changes. From the time the female is about to give birth until 2 weeks after breeding.

FOOD

• Rodent feed. It is also possible to make a homemade mixture of cereals, dried fruits, seeds with almonds, nuts, leftover bread.

• During the breeding season, we have observed that they appreciate small pieces of meat fat, bacon...

• One of his favourites: peanut butter.

• Every day green leaves, lettuce or chard or apple.

WATER

• The most suitable form is the drinking fountains that can be bought in the shops.

shops dedicated to animals.

• We must ensure that the cage is level, otherwise water may drip continuously, soaking the paper and sawdust, and leaving the animals without water.

ENVIRONMENT

• They are nocturnal animals, so a dark place is favourable for breeding, temperatures

pleasant and not extreme.

• During the winter we have to make sure they have enough paper, sawdust, straw to make a nest, and shelter places to keep warm.

• During the summer, not exposed to high temperatures.

HAPPY MICE

They are intelligent, social and curious animals. To enrich their environment and quality of life, we can use a pine cone, toilet paper cartons that act as tunnels, leaf litter or dry leaves and a glass jar without the lid.

BREEDING TIME

• Females are ready for breeding at 6 weeks and males at 8 weeks.

• An adult female is capable of breeding 5 to 10 times a year, producing between 3 and 10 broods at a time.

BIRTH

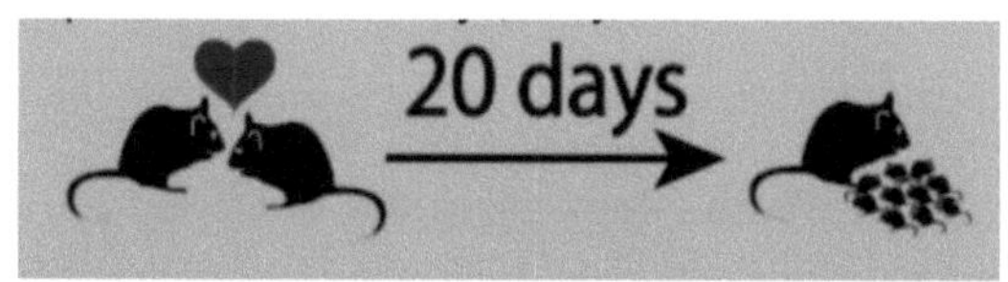

• Pregnancy lasts 20 to 21 days. When the female's pregnancy becomes evident by the size of the belly, as well as after 20 days after birth, it is best not to clean the cage or change the litter of the female.

The problem can cause stress to the parents, resulting in the loss of the pregnancy or the offspring.

• The number of offspring can be between 3 and 10 at a time.

SEPARATION OF OFFSPRING

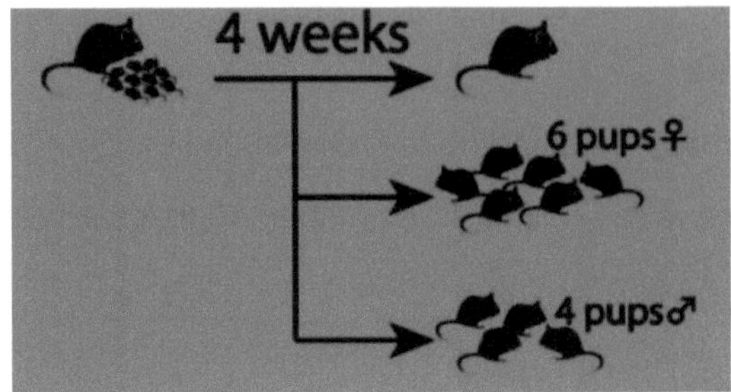

• When the young birds are 21 to 28 days old, they should be separated from their parents, as well as separating the females from the males and keeping them separated in two cages, until the females are 6 weeks old and the males 8 weeks old, when they will be ready for breeding.

• We want to avoid that immature females can become pregnant, since which can affect their health and fertility.

• Males of the same brood do not usually fight, but it is better not to put adult males together because they tend to fight and injure each other.

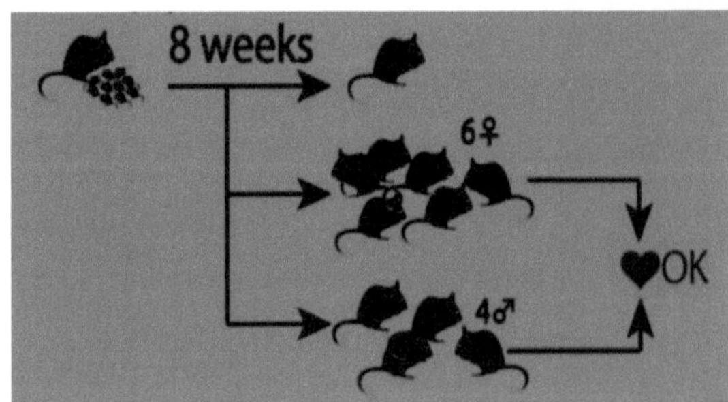

HOW TO DETERMINE THE SEX OF MICE

It is not always easy...

The 2 reference points are the anus and the genitalia, male or female.

•The genitalia of males are placed further away from the anus than the genitalia of females.

• Females have more visible nipples, males do not.

• There is a prominent straight line between the anus and the female genitalia,

while in males it is covered with hair.

Hold the mouse by the tail closest to the body, and let the mouse rest on a flat surface with its front and hind legs supported.

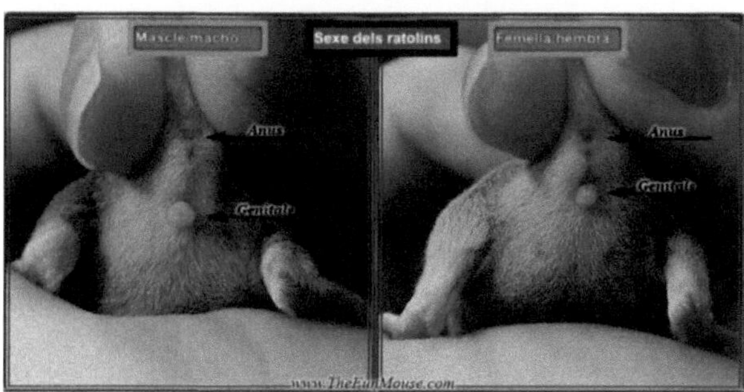

BIRTH CONTROL

It is important to record data

• **Which ones?** date of births, of which female (if possible), number and sex of offspring and date of separation of parents.

THE MOUSE: ITS MICROENVIRONMENT AND MACROENVIRONMENT

Bed Odour Ventilation

Light Density Water

Cage Temperature Feed Noise

The investigator may decide, according to his/her needs, where to place the animals used, taking into account that the site should provide optimal environmental and handling conditions to ensure the health and comfort of the specimens, so that their metabolic and behavioural patterns remain normal and stable, giving reliable responses.

The main environmental factors affecting animals can be classified as follows:

' Climatic: temperature, humidity, ventilation, etc.

Physico-chemical: lighting, noise, air composition, sanitisers, bedding, etc.

Housing: shape, size, type and population of cages, etc.

Nutritional: diet, water, schedule of administration.

Micro-organisms and parasites.

' Experimental situation.

2.1. MICROENVIRONMENT

The microenvironment is the immediate physical environment surrounding the mouse, also called confinement or primary enclosure, and is limited by the perimeter of the cage or box, bedding, food and drinking water; it should contribute to the health of the animals, and avoid any stress, so that each animal should be assigned an adequate space that allows it to move and adopt normal postures, while preserving the minimum conditions of hygiene and protection against insects, rodents and other pests.

The microenvironment consists of the box or cage, food, water, as well as the maintenance of hygienic conditions for each of them (chapter on care and maintenance).

2.1.1. BOX OR CAGE

The mice are housed in specially designed boxes or cages designed to facilitate their welfare, which may be made of metal or plastic (polypropylene, polycarbonate, polystyrene and polysulphan), fitted with stainless steel lids with or without filters.

Polystyrene is transparent and resistant to autoclaving and most disinfectants. Polystyrene and polypropylene do not withstand high temperatures. The height of the box walls should not be less than 12.7 cm.

It must have the following characteristics:

Provide adequate space, be enclosed, secure and protected from external threats.

Be adequate in ventilation.

Be resistant to frequent washing, disinfection and sterilisation.

Allow observation of the animal.

Have easy-to-clean floors and walls (smooth surfaces) and removable grilled or perforated lids.

Maintain in good condition for use.

' Facilitate animals' access to water and feed.

' No sharp edges or projections likely to cause injury.

2.1.2. SPACE RECOMMENDATIONS (ANIMAL DENSITY

The number of animals per cage shall be in relation to the body size (age of the mouse, pre- and postnatal status) and overloading shall be avoided. The size of cages or boxes should be appropriate; for example, in the case of adult mice, a minimum area of 80 cm2 per animal is required. The minimum requirement is that the animal should have sufficient space to move around and to express normal behavioural and social postures, should have easy access to water and food, and should have a sufficient area with clean, unobstructed bedding material to move around and rest.

The following table shows space allowance recommendations for rodents housed in groups; if housed individually or exceeding the weights listed, they may require more space, according to professional judgement and experience.

2.1.3. BED OR BED

The litter shall be made of absorbent material such as wood shavings, ground corn stover, etc.; free of dust, allergens and other substances. toxic. They must be sterilisable. The most suitable wood shavings are white pine shavings, followed by screw shavings.

Chip quality specifications must be available for procurement, such as:

' Not to be harmful.

Absorption capacity

The use of cedar or mahogany shavings is not recommended.

2.1.4. DRINKING WATER

Water should be potable and freely supplied throughout the life of the animal, either in glass or polycarbonate drinking bottles. Water should be acidified, sterilised by autoclaving or by filtration method.

2.1.5. FOOD: DIETS AND REQUIREMENTS

Food is the primary material from which body tissues and structures are formed and renewed, both new and existing ones, which must be replaced due to the process of wear and tear. Nutrition is a determining factor in the successive stages of growth and production of animals, which is why there are specific feeds for each species and even for each stage of their lives. After purchase, care must be taken in the transport, storage and handling of feed to minimise the introduction of diseases, parasites and potential disease vectors (e.g. insects and other pests) and chemical contaminants.

A procedure should be in place for the procurement of feed and the requirements it must meet, such as:

Composition, which shall cover the growth, gestation, lactation and maintenance needs of the mouse.

It must be palatable and digestible.

' Have a date of manufacture and expiry.

Certificate of proximate chemical and microbiological analysis for each batch.

Be free of fishmeal, additives, drugs, hormones, antibiotics,

pesticides and pathogenic contaminants.

The pelleted feed must have the required consistency, to avoid feed loss and to allow the animal to consume the feed.

TABLE 1 Chemical composition of a standard diet

Component	Percentage
Crude protein	20
Crude fat	9,81
Crude fibre	2,15
Ashes	6,38
Daily food consumption	3-6 g
Daily water consumption	3-7mL

2.2. MACRO-ENVIRONMENT

The macroenvironment is the space immediately adjacent to the microenvironment and is the housing room in its general scope. Alteration of the macroenvironment factors will result in changes to the animal model and with it, modification of the type of response, and increased variability of results between or within experimental laboratories.

2.2.1. AIR AND VENTILATION

Rooms used for the production of animals must be ventilated inside with positive air pressure with respect to corridors or outside areas, maintaining pressure gradients in such a way as to prevent pathogens from entering from the outside.

In case of a double-aisle biotherium with central rooms (clean and dirty circulation), the pressure gradient shall be from clean to dirty.

Ventilation is important to control humidity, heat, toxic gases. It should generate 15 to 20 air changes / hour.

Air-conditioning or ventilation systems shall not be shared with other areas, shall be exclusive to the bioterio sector and shall have controlled temperature and humidity factors.

2.2.2. TEMPERATURE AND RELATIVE HUMIDITY

Temperature requirements for mice are 20 to 25 °C and relative humidity between 40 and 70%. The environmental conditions in which the animals are reared and tested have a decisive influence on the responses to different treatments. If standardised responses are required, the conditions under which the animals are kept should be fixed.

2.2.3. LIGHT INTENSITY AND TYPE OF LIGHTING

The rearing rooms shall have artificial light, provided by daylight type fluorescent lamps, with oblique incidence, with a maximum illumination of 323 lux at one metre from the floor; in such a way that all cages, regardless of their location, receive similar light intensities. Lighting should be adequately distributed throughout the housing room and be sufficient for maintenance, inspection and welfare practices without causing clinical signs to the animals. It should also provide safe working conditions for staff.Lighting is important for the regulation of the oestrous and reproductive cycle. We recommend 12 hours light/12 hours dark, which is programmed with a timer.

2.2.4. NOISE

Mice are very sensitive to noise and can perceive sound frequencies that are inaudible to humans, so staff should try to minimise the generation of unnecessary noise. Excessive and intermittent noise can be minimised by training staff in alternatives to noise-producing practices. Radios, mobile phones, alarms and other sound generators, even with headphones or earphones, should not be used in animal housing rooms. A maximum noise level of 85 decibels, higher decibels have harmful effects such as stress and fertility problems.

2.2.5 ODOUR

Odour is another factor that affects the mouse, therefore, disinfectants that emanate odours, which are irritating and much less deodorising, should not be used in the biothereum environment. The perception of ammonia in the environment is an indicator of bed saturation, so it is recommended to have bed change programmes according to the population being managed.For example, it is known that humans are able to perceive 100 ppm of ammonia from the mouse's environment and the mouse can perceive as little as 25 ppm of ammonia.

GUINEA PIG BREEDING OR CUYES

Cobaya, a common name for several genera of small rodent mammals native to South America. These include: the guinea pigs or domestic guinea pigs, the cuises or cuis serranos, the rock guinea pigs and the Patagonian hares or maras. Guinea pigs and rock guinea pigs resemble guinea pigs, but with variations in colour and fur. Patagonian hares resemble rabbits, although they have shorter ears and longer hind limbs, and are between 45 and 75 cm long. All guinea pigs have four toes on the front feet and three on the hind feet. Most of them have crepuscular habits (they are active during dawn and dusk), feed on plant matter, dig burrows and live in large groups. The female gives birth to two young after a gestation period of two months; the number of young may be higher in domestic varieties. The young are very precocial and, despite a period of lactation, are able to eat solid food within a few days of birth. Rock guinea pigs are distributed in northeastern Brazil and live in arid and rocky terrain. Maras are distributed in central and southern Argentina, and inhabit arid, almost desert-like regions.

Scientific classification: guinea pigs belong to the family Cavidae, in the order Rodents. Guinea pigs are classified in the genus Cavia. Guinea pigs are classified in the genera Cavia and Galea, and rock guinea pigs in the genus Kerodon. Finally, Patagonian hares are classified in the genus Dolichotis.

Cavia porcellus is a hybrid domestic species of histricomorphic rodent of the family Caviidae, resulting from the crossing of several species of the genus Cavia in the Andean region of South America, with archaeological records found from Colombia and Ecuador to Peru and Bolivia. It weighs up to 1 kg. It lives between five and eight years. The species was first described by the Swiss naturalist Conrad von Gesner in 1554.[1] Its scientific name is due to Erxleben's description in 1777, and is a mixture of the genus designation by Pallas (1766) and the specific name given by Linnaeus (1758).[2]

Common names around the world

In Spanish, Cavia porcellus has different common names depending on the country. In its area of origin (Colombia, Ecuador and Peru) it is known as **cuy** (from Quechua quwi), an onomatopoeic name that it still bears in some regions of South America. Mainly in South America, but also in Mexico and Central America, there are several forms of the onomatopoeic Quechua name quwi: **cuye**, **cuyi**, **cuyo**, **cuilo**, **cuis**. In countries of the Caribbean area, Andalusia and the Canary Islands, the name has derived to **curi**, **acure**, **curí**, **curío**, **curie**, **cury**, **cuín** and **curiel**.

In Chile, it is generally known as **cuyi**, but also in its northern areas it is known as **cuy**, due to the ancient Quechua presence and later Peruvian immigration to Chile. In Santiago de Chile it is called **cuyi**, **cuy**, **cuye** or **cuyo**.

In the highlands of the Peruvian department of Huánuco, it is known as jaka (from the Quechua haka).

In Uruguay it is known as "cuis".

In Spain and parts of Latin America, the names **cobayo** and **cobaya** are used,[3] possibly derived from the Tupi language sabúia. In many countries, including those already mentioned, it is called **conejillo de indias (guinea pig)**, in the Rio de la Plata region it is called **chanchito**

de indias,[4] while in the rest of Argentina the word cuis or cobayo is used. In Puerto Rico the name **güimo** is commonly used.

Etymology in other languages

The name given to the species Cavia porcellus in other European languages is completely unrelated to the original.

• cavia peruviana or porcellino d'India ('little Indian pig') in Italian.

• porquinho da Índia ('Indian piglet'), in Portuguese

• cochon d'Inde ('guinea pig') or cobaye in French

• guinea pig', in English

• Meerschweinchen (German for 'little sea pig')

• морская свинка or morskáia svinka ('sea piglet'), in Russian

The origin of all these names is difficult to explain, although there is one hypothesis:[cita required] perhaps the German and English traders who brought it to Europe were returning by sea from Guinea, which could have confused the origin of the animal. Another hypothesis is that the name of the animals could be related to the coin "guinea", an English gold coin.

Relationship with the human being

Its domestication for human consumption took place 2500 years ago in the Central Andes, specifically in the department of Junín (Peru), in the same region where the domestication of alpacas took place.[5] Throughout time, Andean man has raised guinea pigs for their meat and even in some areas to make clothes with their skin; a clear example is found in the Ecuadorian highlands. In the Andean countries there is a stable population of more or less 35 million guinea pigs, Peru being the country with the highest consumption and population of guinea pigs, with an annual consumption of more than 65 million guinea pigs, produced by a more or less stable population of 22 million animals raised basically in family production systems.[6] The estimated self-consumption population in Ecuador is 15 million heads of guinea pigs,[7] , which is much lower than the commercial production, estimated at 50 million. Another reason for breeding this rodent is to market it as a pet.

Research use

The guinea pig is a very common animal for experimentation in biomedical research, hence the expression guinea pig is popularly used as a synonym for experimental object.

As a pet

Nowadays it is increasingly bred as a pet, as it is able to live with small children. As such, the so-called type 3 guinea pig is preferred, i.e. the long-haired, straight-haired guinea pigs known as "Angora guinea pigs". Although some short-haired species are also preferred as pets.

Feeding

Guinea pigs are herbivorous animals, so fibre in their food is indispensable. On the other hand, the supply of vitamin C is highly necessary, as guinea pigs, primates and bats are the only species that do not synthesise this vitamin, and if they are not provided with vitamin C, it is essential for them to have a high intake of it. They can develop scurvy and die, so pepper, orange and guava should be included in their diet.[8] A well-balanced diet should consist of fresh vegetables, hay, water, supplemented with kibble or commercial food. To prevent deficiencies, the pet should be provided with a varied diet. Hay serves to meet the pet's need for carbohydrates and fibre. Alfalfa provides calcium for their bones and is essential. Fruit and vegetables help to meet their vitamin and fluid requirements. For food, it is best to use heavy earthenware bowls that will resist tipping and consequent dropping of the food. Their sides should be high enough to keep bedding and faeces away from the food. It is also very important that all fresh food given to guinea pigs is at room temperature; it should never be fresh from the refrigerator.

Much of their fluid needs are met by the ingestion of fresh food. A drinking fountain with clean, fresh water should be available at all times. If water bottles equipped with a drinking tube are used, it will be easier to keep the water free of contamination. Guinea pigs tend to contaminate and clog their water bottles more than other domestic rodents, as they chew on the tube in order to obtain the water, introducing food particles into the bottle. For these reasons, all food, and water containers in particular, should be cleaned regularly.

As herbivores, rabbits should not be given meat, dairy products or rabbit pellets (they do not contain vitamin C and some may even include antibiotics that are toxic to guinea pigs).[9]

Guinea pigs are animals that perform cecotrophy, a form of coprophagy specific to certain rodents; that is, they eat faeces directly from the anus, before they reach the ground.[cita required] This is a good way of taking advantage of all those nutrients that have passed directly through the gastrointestinal tract without having been absorbed, such as some vitamins, for example. Guinea pigs do not digest nutrients well on their first pass through the digestive tract, so they eat their faeces to pass them through the stomach again. Also to prevent overgrowth of their teeth, they should be provided with food that wears down their teeth, such as hay and corn husks and also some wooden sticks for them to gnaw on.[10]

Health and hygiene

In order to keep healthy guinea pigs and avoid diseases you must:
• Feed them well.

• Keep cages clean.

• Avoid the presence of spoiled food.

• Newly acquired animals   should be quarantined for four days to observe their behaviour before being placed with existing animals. The introduction of a new animal should always be done in a neutral, odour-free territory to facilitate integration.

A healthy guinea pig is a happy animal, with a shiny coat, plump, well-developed and a good eater. A guinea pig is sick when it separates from the others, withdraws into a corner, is listless, does not want to eat, its fur stands on end, its belly sinks, it has diarrhoea and loses weight rapidly. In this case, it is necessary to separate it from the others quickly so that it does not infect them and to go to a vet specialised in exotic animals.

The most common diseases of guinea pigs are the following:
• Infection with external parasites: lice, fleas, ticks and mange. This can be controlled by good cage hygiene.
• Cecal dysbiosis: this is very serious because it can lead to the death of the animals. It can be caused by various factors, such as bacteria

(Clostridium piriforme), low fibre levels and excessive carbohydrates of easy fermentation, which generate cecal hypomotility. Regular hygiene and disinfection of cages is recommended.

• Pneumonia: use specific antibiotics and avoid cold and draughts

• Scurvy: this is caused by a lack of vitamin C and leads to internal haemorrhages. In this case, 2 drops of Redoxon (vitamin C drops) should be administered per 100 g of body weight. The treatment should last until the guinea pig gets better.

There are two main breeds of guinea pigs for food, as well as several lines:[citation needed]

Race Peru

It is characterised by good meat conformation, precociousness (i.e. rapid growth or fattening) and low prolificacy. Its colours are red and white.

Andean Breed

Characterised by good conformation and prolificacy, but less precocious than the Peruvian breed. They are pure white with black eyes.

Inti Line

It is characterised by being an average of the two previous breeds. It is a more forage animal and its colours are yellow or bay with white.

BREEDING OF HAMSTERS OR CRICETUSES

Hamster, a common name applied to several species of mammalian rodents. They are characterised by having pouches on each cheek, also called abaxons, which they use to store food, by having dense, soft fur and a robust body, with short legs and tail. One species can be about 30 centimetres long. Hamsters live in underground burrows consisting of several chambers, one of which is used for food storage, consisting mainly of cereal grains. Sometimes these animals are serious agricultural pests. Hamsters hibernate during the winter, but often wake up to eat the stored food. Another chamber of the burrow is used as a nesting chamber, where the female hamster gives birth to a litter of up to 18 offspring several times a year. The young hamsters are weaned after two to three weeks, and shortly afterwards they abandon their parents and build their own burrows. In some places their meat is eaten, and the fur of some species is used as coat lining.

Hamsters range throughout Eurasia and across the steppes, plains and deserts of central Asia. The golden hamster is a highly prized species, as it is used in medical research and as a pet. In 1956, it was discovered that the hamster could also contract the common cold disease. This discovery was of great importance in research into the virus that causes it in humans.

Scientific classification: Hamsters belong to the family Muridae (Muridae), in the order Rodents. The common hamster is scientifically named Cricetus cricetus, golden hamsters are classified in the genus Mesocricetus, and the dwarf hamster belongs to the genus Phodopus.

BREEDING OF SQUIRRELS

Squirrels belong to the family Scyuridae, in the order Rodents. American tree squirrels, ground squirrels, woodchucks and chipmunks belong to the subfamily Scyuridae, and flying squirrels belong to the subfamily Petauristinae. The African pygmy squirrels constitute the genus Myosciurus and the Asian giant squirrels the genus Ratufa. The common European or red squirrel is classified as Sciurus vulgaris; and the American or grey squirrel as Sciurus carolinensis. The American pygmy squirrels form the genera Microsciurus, Sciurillus and Guerlinguetus, among others. The smaller ones, the genus Leptosciurus; the medium-sized ones, the Neosciurus; and the larger ones, the Hadrosciurus. Scaly-tailed squirrels belong to the family Anomaluridae.Squirrel, a common name applied to certain species of rodents belonging to a family that includes tree squirrels, ground squirrels, groundhogs, woodchucks, chipmunks, flying or gliding squirrels and prairie dogs. The tree and ground squirrel group includes about 230 species and the flying squirrel group includes about 43 species. They vary in size; the smallest are the African pygmy squirrels, which are about 13 cm long, and the largest are the Asian giant squirrels, which are about 90 cm long. Squirrels are distributed throughout the world, except Australia. They usually live in deciduous or coniferous forests, although it is possible to find species adapted to live in very different habitats, from taiga to desert. To live in these places, squirrels have adapted and developed strategies that allow them to withstand the extreme temperatures that characterise them. With the exception of ground squirrels, most species are arboreal and feed on plant matter (mostly nuts, seeds and sprouts), although they may occasionally eat insects. Their habit of storing food for the winter helps the propagation of plants and trees. Squirrels are very sensitive to temperature changes, remaining inactive in warm climates when temperatures are high, and those living in colder climates hibernate, although tree squirrels never do. The common or European squirrel is reddish, has a long tail and small ear tufts. The grey squirrel is larger than the European squirrel and has no ear tufts. The arboreal species common in Latin America are classified into four main groups: the pygmy and guerligueto squirrels, the lesser squirrels of South America, the medium-sized squirrels of the north and the West Indies (also present in Central America) and the greater squirrels of South America.

The **red squirrel (Sciurus vulgaris)**, or simply the **common squirrel**, is a species of scyuromorph rodent of the family Sciuridae. It is one of the most widespread squirrels in the forests of Europe.

Features

Its body measures between 20 and 30 cm and its tail between 15 and 25 cm. It weighs 250 to 340 g. Its coat is reddish in colour. When winter arrives, tufts of hair appear on its ears. Its front legs or hands have four toes, while its hind legs have five. There is no sexual dimorphism. Their tooth formula is as follows: 1/1, 0/0, 2/1, 3/3 = 22.[2]

Behaviour

It is a common inhabitant of coniferous forests, although it is also present in other tree formations. It is active during the day, searching for and consuming fruit, seeds, bark and even insects, eggs and birds. It does not hibernate but remains active by consuming what it has stored on the ground and in different hollows in trees and rocks. It is active in the trees, although it does not hesitate to climb down to gather food. It also swims with ease, using its tail as a rudder to change the direction of its movement. When it sees an enemy, it wags its tail and makes loud noises so that the rest of its relatives know about it.

Reproduction and mortality

The breeding period occurs in late winter and summer. A female has two litters per year, usually with three or four pups, exceptionally six. Gestation lasts 38-39 days. They are born helpless, blind, deaf, weighing 10-15 g; their body is covered with fur at 21 days, eyes and ears open after 3-4 weeks, they develop teeth at 42 days. They begin to eat solid food at 40 days and weaning occurs at 8-10 weeks. Males detect females in oestrus by their scent, and although there is no courtship and multiple males advance to a single fertile female, eventually the dominant male, usually the largest in the group, joins her. Males and females mate multiple times with many mates. The female must reach a minimum body weight before she goes into oestrus and the heavier female produces more offspring. If food is scarce, pregnancy may be lost. Typically, a female produces her first litter in the second year. Life expectancy is, on average, three years, although it can reach 7, and 10 in captivity. Survival is positively linked to seed availability in autumn-winter. 75-85 % of juveniles die during their first winter, and mortality is about 50 % for subsequent winters.[3]

Taxonomy

Up to 40 subspecies have been named, although the taxonomic status of many is uncertain. A 1971 study recognised 16 subspecies and served as a basis for subsequent taxonomic work.[4][5] Currently, the number of recognised subspecies is 23.[6]

Scientific classification: Squirrels belong to the family Scyuridae, in the order Rodents. Tree squirrels, ground squirrels, marmots and American chipmunks belong to the subfamily Scyuridae, and flying squirrels belong to the subfamily Petauristinae. The African pygmy squirrels constitute the genus Myosciurus and the Asian giant squirrels the genus Ratufa. The common European or red squirrel is classified as Sciurus vulgaris; and the American or grey squirrel as Sciurus carolinensis. The American pygmy squirrels form the genera Microsciurus, Sciurillus and Guerlinguetus, among others. The smaller ones, the genus Leptosciurus; the medium-sized ones, the Neosciurus; and the larger ones, the Hadrosciurus.

BREEDING OF HEMP

Jerbo, common name for certain jumping rodents that live in the arid regions of North Africa and Central Asia. They are small to medium-sized animals and measure between 4 and 26 cm, not including the tail. They are characterised by short forelimbs, especially in comparison with the hind limbs, which are very long (10 cm in length), as they are adapted for jumping. The tail is also long, with a tuft of hair at the end, and is often used for balance when jumping or as a support when sitting. The ears and eyes are large, the fur is soft and usually sandy on top and white underneath. They inhabit arid regions with sparse, sparse vegetation; they live in burrows they dig in the ground and feed on plants, seeds and insects. They are very well adapted to their environment, as they are able to go for long periods without drinking; they obtain the water they need from the food they eat. Some species breed once a year, others twice a year, and the litter usually has an average of three young.

Scientific classification: Gerbils belong to the family Dipodinae.

Dipodines (Dipodinae) are a subfamily of myomorph rodents of the family Dipodidae commonly known as **gerbils** (from ‏ربوع‎ Arabicyarbū' or Hebrewyarbōayٳ‏ירבוע‎ '). They are jumping rodents that live in northern

Africa and Asia. They are not to be confused with gerbils (such as Meriones unguiculatus), which belong to the family Muridae. There are nine different species of gerbils, distributed in five genera,[1] the most common and used as a pet in several countries is the Egyptian gerbil (Jaculus jaculus). Unlike gerbils there are 16 genera, and more than 110 different species.The gerbil is often confused, because of the similarity of their names, with the gerbil, but they are very different species, from different families, and the only thing they share is that they belong to the same order of rodents. There is no single species called a gerbil, but Jaculus jaculus is generally known as the gerbil because it is the most sought after pet in countries such as the United States, Canada and also in Europe.

Features

Gerbils are similar to rats, but differ from rats in that their hind legs are used for hopping, similar to those of a kangaroo; their front legs are not used for movement. They feed mainly on seeds, insects and vegetables. African gerbils have 3 toes on their legs, unlike the Asian gerbil, which has 5 toes on its legs. Gerbils have managed to develop a defence system using long jumps to escape from predators, considering that their natural habitat is desert. Gerbils have become popular as pets due to their longevity among rodents and their docile, playful, peaceful and extremely curious nature.

Taxonomy

There are five genera and nine species of gerbils:[1]
• Dipus
  ○ Dipus sagitta

- Eremodipus
  - Eremodipus lichtensteini
- Jaculus
  - Jaculus blanfordi - Blanford gerbil
  - Jaculus jaculus - Egyptian gerbil
  - Jaculus orientalis - great eastern gerbil
  - Jaculus bishoylus - great Egyptian gerbil
- Paradipus
  - Paradipus ctenodactylus
- Stylodipus
  - Stylodipus andrewsi
  - Stylodipus sungorus
  - Stylodipus telum

Endangered

There are two species that are considered threatened: the pygmy five-toed (classified vulnerable), and the pygmy thick-tailed (classified vulnerable). Many other species have been placed in a "lower risk" category.

CHINCHILLA BREEDING

Chinchilla, a common name for certain species of rodents that inhabit the Andes mountain range in South America at altitudes between 3,000 and 5,000 metres. These mammals are reminiscent of squirrels in appearance; they have rather long ears, large eyes, short legs, long whiskers and a furry tail. They measure between 23 and 28 cm in length, not including the tail, which is between 7 and 15 cm long. They live in colonies and usually seek shelter in holes between rocks and boulders. Like their relatives the vizcachas, they feed on the hard vegetation in their habitat. It is common to see them clutching the roots and grasses they eat with their forelegs. The hind legs are proportionally longer and enable the animal to leap nimbly. Females give birth to an average of two to three young per litter and can reproduce twice in the same year. The young are precocious: they run within hours of birth and eat solid food within a few days.

Chinchillas are covered with soft, dense fur; it is silvery grey on the back and whitish on the belly. Their fur is highly prized and consequently they have been hunted to near extermination. During the 1920s, laws were adopted for its protection and breeding farms have been established ever since.

Scientific classification: Chinchillas belong to the Chinchillidae family (Chinchillidae), in the order of Rodents. The chinchilla real or indiana is the species Chinchilla chinchilla chinchilla, is the largest and lives in Peru; the chinchilla costina of Chile is the species Chinchilla lanigera, and the chinchilla del altiplano, species Chinchida intermedia, lives in certain localities of Argentina and Bolivia.

BREEDING OF HUTIA

Capromyinae

Jutías

Time range: Lower Miocene – Recent

Capromys pilorides

Taxonomy

Kingdom: Animalia

Phylum: Chordata

Class: Mammalia

Order: Rodentia

Suborder: Hystricomorpha

Family: Echimyidae

Subfamily: CapromyinaeSMITH, 1842

Genres

Capromyids (Capromyinae) are a subfamily of rodents of the family Echimyidae commonly known as **hutias** that inhabit the Caribbean.[1] They are similar to the Cavia. Twenty species are known and half of them are at risk of extinction.

Features

They resemble coypus or otters in several respects and the largest species reach several kilograms in weight. They have vestigial to prehensile tails. They have robust bodies and large heads. Many species are herbivores, although some eat small animals. Instead of digging caves, they make nests in trees or rock cavities. They are hunted for their meat in Cuba, where they are often cooked in large pots with wild hazelnuts and honey. One species of hutia is referred to at the Guantanamo Naval Base as "Banana rat" by the military stationed there.

Ranking

The following species are recognised:[2][1]

Tribe Capromyini Smith, 1842
• **Genus** Capromys
  ○ Capromys pilorides - Conga hutia
• **Genus** Geocapromys
  ○ Geocapromys brownii
  ○ Geocapromys ingrahami - Bahama hutia
  ○ Geocapromys thoracatus †
• **Genus** Mesocapromys
  ○ Mesocapromys angelcabrerai
  ○ Mesocapromys auritus - Rat hutia
  ○ Mesocapromys nanus - Dwarf hutia
  ○ Mesocapromys sanfelipensis - ground hutia
• **Genus** Mysateles

○Mysateles melanurus[3] - Houbara bustard

○Mysateles garridoi - Jutía de Garrido (This species does not appear in the list of native species because it is considered a taxon of dubious identity (species inquerenda) in the recent study by Silva et al. (2007).[cita required)]

○Mysateles prehensilis - Carabali hutia

Tribe Hexolobodontini † Woods, 1989

• **Genus** Hexolobodon †

○Hexolobodon phenax †

Tribe Isolobodontini Woods, 1989

• **Genus** Isolobodon

○Isolobodon montanus †

○Isolobodon portoricensis†[4]

Tribe Plagiodontini Ellerman, 1940

• **Genus** Plagiodontia

○Plagiodontia ipnaeum† - Samana hutia

○Plagiodontia aedium - Hispaniolan hutia

○Plagiodontia araeum†

• **Genus** Rhizoplagiodontia†

○Rhizoplagiodontia lemkei†

REFERENCES

• National Biodiversity Centre, Cuba (17 October). "Cuban Biological Diversity". Archived from the original on 19 December 2009. Accessed 25 May 2010.

1. 1 2 Fabre, Pierre-Henri; Upham, Nathan S.; Emmons, Louise H.; Justy, Fabienne; Leite, Yuri L. R.; Loss, Ana Carolina; Orlando, Ludovic; Tilak, Marie-Ka; Patterson, Bruce D.; Douzery, Emmanuel J. P. (1 March 2017). "Mitogenomic Phylogeny, Diversification, and Biogeography of South American Spiny Rats". Molecular Biology and Evolution **34** (3): 613-633. ISSN 0737-4038. doi:10.1093/molbev/msw261.
2. ↑ Wilson, D. E. & Reeder, D. M. (editors). 2005. Mammal Species of the World. A Taxonomic and Geographic Reference (3rd ed).
3. ↑ Soy, J. & Silva, G. 2008. Mysateles melanurus. In: IUCN 2010. IUCN Red List of Threatened Species. Version 2010.1. Last accessed 26 June 2010.
4. ↑ Turvey, S. & Dávalos, L. (2008). "Isolobodon portoricensis". IUCN Red List of Threatened Species 2008. Accessed 6 January 2009.

BIBLIOGRAPHY

• Alvarado, R. 1970. Los Roedores. Editorial Noger, S. A. Barcelona-Madrid, 80 pp.

• Hernández-Muñoz, A .2024. Allies of Man. Editorial Académica Española, 126 pp.

• Hernández-Muñoz, A. 1995. Cuyicultura. Forum de Ciencia y Técnica. Sancti Spíritus, 43 pp.

• Laurent, O. 1998. The Mice. Editorial Vecchi. Barcelona, 95 pp.

• Varona, L. S. 1980. Mammals of Cuba. Editorial Gente Nueva. Havana, 109 pp.

• Varona, L. S. 2007. Mammals of Cuba. Editorial Gente Nueva. Havana, 134 pp.

• "Common Squirrel." Microsoft Encarta. 2009. [DVD]. Microsoft Corporation, 2008.

• "Cobaya." Microsoft Encarta. 2009. [DVD]. Microsoft Corporation, 2008.

• "Chinchilla." Microsoft. Encarta. 2009. [DVD]. Microsoft Corporation, 2008.

• "Hamster." Microsoft Encarta. 2009. [DVD]. Microsoft Corporation, 2008.

• "Jerbo." Microsoft Encarta. 2009. [DVD]. Microsoft Corporation, 2008.

• http//www.wikipedia.com

• http//www.wikiespecies.com

I want morebooks!

Buy your books fast and straightforward online - at one of world's fastest growing online book stores! Environmentally sound due to Print-on-Demand technologies.

Buy your books online at
www.morebooks.shop

Kaufen Sie Ihre Bücher schnell und unkompliziert online – auf einer der am schnellsten wachsenden Buchhandelsplattformen weltweit! Dank Print-On-Demand umwelt- und ressourcenschonend produziert.

Bücher schneller online kaufen
www.morebooks.shop

Printed by Books on Demand GmbH, Norderstedt / Germany